YOUR KNOWLEDGE HAS VALUE

- We will publish your bachelor's and master's thesis, essays and papers

- Your own eBook and book - sold worldwide in all relevant shops

- Earn money with each sale

Upload your text at www.GRIN.com and publish for free

Cow Milk Production, Practices and Marketing System of Milk and Milk Products in Sodo Zuria District, Wolaita Zone, Southern Ethiopia

Mesfin Getachew Keshamo

Bibliographic information published by the German National Library:

The German National Library lists this publication in the National Bibliography; detailed bibliographic data are available on the Internet at http://dnb.dnb.de.

ISBN: 9783346337153
This book is also available as an ebook.

Assessment of Cow milk Production Practices and Marketing System of milk and milk products in Sodo Zuria District, Wolaita Zone, Southern Ethiopia

Mesfin Getachew

1. Wolaita Soddo ATVET College, Department of Animal Production, Ethiopia

Abstract

This study was conducted to assess the cow milk production practices and marketing system of cow milk and milk products in Sodo Zuria District, Wolaita Zone, and Southern Ethiopia. To conduct the study, one sample district (Sodo zuria) was purposively selected based on dairy cow potentiality and milk products marketing. 162 dairy producers' households both highland and midland producers in the six major kebeles representing the Sodo Zuria district area in Wolaita Zone, Southern Ethiopia, were selected using a multi-stage sampling techniques, questionnaire based formal survey were employed to collect both quantitative and qualitative data on cattle milk production practices in the district and a Rapid Market Appraisal (RMA) techniques were used to assess the marketing system of milk and milk products. Two main dairy production systems; mixed crop-livestock production system (78.9%) and urban dairy production (21.1%) were identified. In the study district, local dairy cows (88.8%) and HF/Jersey crossbred cows (11.2%) raised by smallholder farmers. The average cattle herd size of households was 1-3 heads. From local cattle breeds, cows (2.8%) and calves (3%) mainly dominate the herd composition, while heifers (1.9%) and oxen (1.3%) represented small percentage. Average milk off-take from indigenous and their crosses (HF or Jersey) cows was 1.4±0.22 and 6.43±0.34 litre/head/day, respectively. The overall mean milk off take from local cows at first, second, and third lactation stages were 2.23±2.85 litre, 1.52±2.28 litres and 0.72±1.23 litre, respectively while for crossbred cows it was 4.53±0.621, 7.06±1.72 and 4.43±0.85, respectively. The overall average lactation length was 8±0.19 months for local cows, while it was 9±0.57 months for crossbreed cows. Overall mean CI was 16.15±5.1 months for local cows, while it was 20.01±5.53 months for crossbreed cows. Mean AFC for local cows 4.15 years, while their counter crosses had 3.12 years. In general, the market share of whole milk and other milk derivatives (fermented milk) was almost negligible, while butter was comparatively the most marketable commodity in study area. Lack of fodder, low milk yield, poor quality of fodder/roughage and high price of concentrate feed were mentioned by farmers as major constraints of milk production. The rapid urbanization of Sodo town, subsequent increase in human population and the willing of farmers can be considered as good milk production opportunities in the study area. Dairying can be improved by solving constraints such as feed supply, veterinary care, and milk processing facilities, AI services, and extension services and developing efficient marketing systems.

Keywords: Agro-ecology, Breed type, marketing, milk production.

Table of content

1. Introduction

Ethiopia possesses the largest livestock population in Africa. Estimates for farmer holding in rural areas indicate that the country has about 60.39 million heads of cattle, 32.74 million goats, 31.30 million sheep, 2.01 million horses, 8.85 million donkeys, 0.46 million mules, and 1.42 million camels (CSA, 2017/18).

In Ethiopia, the traditional milk production system and marketing which are dominated by local breeds of low genetic potential for milk production, and accounts for about huge number of the country's total annual milk production (Abraha, 2018). Demand for milk and milk products exceed supply that is expected to increase growth in dairy sector (Haese *et al.*, 2007).

Milk production systems in Ethiopia is classified into urban, per-urban, and rural (Reda, 2001). Livestock rising always has been mainly a subsistence activity. They are raised in all of the farming systems of Ethiopia by crop/livestock mixed farmers.Livestock play a vital role in economic development, chiefly as societies change from subsistence agriculture in to cash-based economies (Agajie *et al.*, 2002). The contribution of the livestock sector is estimated to be about 12% to 16% of the total growth domestic product (TGDP). Only milk and milk products contributed 36%-46% to the house hold income in some areas (Ahmed *et al.*, 2003; Asrat *et al.*, 2013). In Ethiopia, different type of milk production system can be recognized based on various criteria. Milk production systems can broadly categorized in to urban, peri-urban and rural milk production systems based on location (Redda, 2001),while based on market location, scale and production intensity,dairy cattle production system can be categorized as traditional small holders, privatized state farms and urban and peri-urban system (Ahmed *et al.*,2004, Ketema, 2008 and Kumsa, 2002). Urban and peri-urban dairy production system is among the forms of dairy production in the tropics and sub-tropics. The system contains the production and marketing of milk and milk products in the urban centres (O'Connor, 1990). Being of the urban and peri-urban dairy farming is mainly interested by availability of good market for animal products, need for creation of employment opportunities (Prain *et al.*, 2010 and RLDC, 2009). Currently demand for dairy products in the country exceeds supply, which is expected to persuade rapid growth in the dairy sector (Haese *et al.*, 2007, Ketema and Redda, 2004).Factors contributing to this include rapid population growth (FAO, 2004), increaseurbanization andpredictable growth in incomes (Asrat *et al.*, 2014). Soddo zuria district is one of the potential areas for milk production in Wolaita Zone with access for large grazing land. But little is known about the existing dairy production system and constraints associated with dairying though dairy process plays a key role for engaged households. Identification of usual problems and describing of the existing dairy production system in the area is a precondition to make any development interventions. Unlike other part of Ethiopia, Wolaita zoneis well known in livestock population; well suited agro-ecology and vegetation cover for many years it is known that there is little information in dairy production system and marketing. Identification of prevailing problems and understanding of the existing dairy production and marketing system in the area is paramount importance to make future improvement interferences. The objectives of the current study were therefore to assess dairy production and marketing systems and to identify constraints and opportunities of dairy production and marketing system in Wolaita zone.

2. Material and Method

The study was conducted at Sodo Zuria District; is located at 390kms far from Addis Ababa, the capital city of Ethiopia, 170 km from Hawassa, the regional capital of southern nation, nationalities of peoples of republic state. The district has two agro-ecological zones such as highland and midland with annual average daily temperature ranges from $15^{o}c$ to $30^{o}c$ from minimum to maximum respectively and annual average rain fall ranges 1200 mm per annum. The altitude of the district ranges from 1950 metre above sea level and it has 6°54'N 37°45'E latitude and 6.900°N 37.750°E longitude and has a total land area of 4,541 km^2. There are 129,660 Cattle, 35873 Sheep, 9179 Goats, 9115 equines, 11,838 Poultry and 7,712 dairy cows in the district (Soddo Zuria District Livestock and Fishery Office, 2010). The production system of the area is mixed crop livestock production system with crop cultivation as primary and livestock as secondary production (SZWAO, 2009). The major crops grown in the area includes, maize, wheat, teff, and enset at high lands is recognized (SZWLFO, 2011).

2.1. Sampling procedure

The District has 24 kebeles and it has two agro-ecological zones. Then the District stratified in to two on the agro-ecological difference as well as differences in the livestock holding per households. Then six kebeles were selected from the total of 24 kebeles in which three kebeles from highland and midland respectively were selected randomly. Finally the household was selected purposively based on milk production potential, owing milking cows and marketing of milk and its products. From midland (74) and highland (88) households were selected from each kebele i.e Bugewanche, Wachigabusha, Wajakero and Dalbowogane, Warazalasho and Zalashasha kebeles repectively and a total of 162 households were used for the study.

2.2. Data collection

Both primary and secondary data were used in the study. Primary data were collects through interview by using semi-structured questionnaires after translated in to Wolaitegna language and discussed with the owners of dairy cattle. The questionnaires assessed the production system, production traits, reproduction traits, constraints, opportunities and milk marketing system and challenges on milk market in the study area. Secondary data on socio-economic characteristics, agricultural production system farming practices and description of the kebeles were collected from Soddozuriaworeda agricultural office.

2.3. Data Analysis

Both the primary and secondary data were analysed by using descriptive statistics such as mean and percentage and summarized and reported by using tables, graphs and charts. For all parameters, independent samples t-test at $P \leq 0.05$ was used to compare the means difference across agro- ecologies by using the following model.

Model 1: the appropriate model used for this study was as follows:

$$Yij = \mu + \alpha i + \varepsilon ij$$

Where;

Yij = measurement or observation due to ith factor

μ = the overall mean

αi =the effect of factor i (agro-ecology)

εij =random error with mean zero (0) and variance σ^2 and for the data analysis for which the following mathematical General Linear Model was utilized to analyse productive and reproductive performances: The model analyses the effect of agro-ecology and genetic group on performance of cow keeping other factors constant.

Model 2: $Yijkno = \mu + Gi + Lj + (G+L)k + eijlo$

Where,

Yijlo = the value of oth individual under ith genetic group and j^{th} location.

μ = the population mean.

Gi = the effect of ith genetic group (i=1, 2)

Lj = the effect of jth location herd (j=1, 2)

G+L= the interaction effect of genetic group and location

eijklo=the random error associated with individual which is randomly and independently distributed with mean zero and variance α^2.

3. RESULTS AND DISCUSSION

3.1. Demographic Characteristics of Respondents

The family size and age group of respondents in the study area are shown in Table 1. Each age group had a significant difference within each agro-ecology (P<0.05). Most (61.4%) of household heads in highland agro-ecology were in the age group of 25-35 years and 62.2% of the head in midland were in the age group of 36-45 years but the overall percentage is 48.1%. The current finding agrees with the report of Getachew and Tadele (2015) who reported the overall mean age of 46.33±12.87 years for cattle producer households in Cencha Woreda, Gamo Gofa Zone of Southern Ethiopia.

Most dominant family size per family 83% and 75.7%for highland and midland agro-ecologies respectively with overall percentage value of 79.6% per family were 3-5 persons. This finding was comparable with the result reported by (Addisu *et al.*, 2016) who reported that average family size was 5.7 persons per family in Gonder town area whereas (Tollossa *et al.*, 2014) reported that higher value in Boranawhere the average family size was 7.76 persons per family.

The gender of respondents in present study indicated that overall in the study area, the majority (77.8%) were male headed and the rest (22.2%) were female headed. In both agro-ecologies the sex of male household were not significantly varied (P>0.05).

As it is presented in Table 1, educational levels of total household members who can read and write were about 36.3% (14.4% for male and 21.9% for female). Among household members, 29.2% were illiterate, 29% were in high school, and 5.5% completed high school education.

Percent illiteracy rate in the current study area is higher with 22.5% illiteracy rate as indicated in the report of Getu *et al.* (2016) in Amhara region, Ethiopia, 27.78% illiteracy rate reported in Bench-Maji Zone of Southwest Ethiopia (Weldegebriel, 2015) and 10% illiteracy rate indicated in the report of Aleta Chukko district of Southern Ethiopia by Beriso *et al.* (2015).

Table 1: General demographic characteristics of respondents in the study Area

Agro-ecology Items	Highland	Midland	Overall	P-value
Household size	%	%	%	
Below 2	1.1	2.7	1.9	
3-5	83.0	75.7	79.6	
6-9	13.6	21.6	17.3	0.265
Above 9	2.3	0	1.2	
Age of HH head				
25 to 35	61.4	29.7	46.9	
36 to 45	36.4	62.2	48.1	0.001
46 to 65	1.1	6.8	3.7	
Above 66	1.1	1.4	1.2	
Sex of HH head				
Male	79.5	75.7	77.8	
Female	20.5	24.3	22.2	
EDU ST				
Male illiterate	39.8	38.9	13.9	0.255
Male grade 1-8	39.8	41.7	14.4	0.094
Male grade 8-12	45.8	48.6	16.7	0.011
Male grade >12	9.6	11.1	3.4	0.248
Female illiterate	47.0	40.3	15.3	0.010
Female grade 1-8	66.3	56.9	21.9	0.001
Female grade 8-12	20.5	51.4	12.3	0.001
Female grade >12	1.2	11.1	2.1	0.080

HH=House hold,EDU ST=Educational Status,The p-values denote significance difference between agro ecologies (p<0.05)

3.2. Socio-economic characteristics of households

The overall average land holding per household of study households were 1.16±0.65ha, 0.32±0.22ha and 0.26±0.49ha for crop production, grazing land own and grazing land rented respectively. Due to rapid urbanization and provision of communal land to unemployed youth in the area, farmers do not have extra land to develop improved animal feeds or do not have access to communal grazing land. As indicated in Table 2, there is small land size in the study area, but compared to the regional average land holdings of Southern Nation, Nationalities of Peoples Republic State (SNNPRS), the overall mean value of 1.16± 0.65ha for this study area is not low compared to the fact that 46.5% of the farmers in SNNPRS households own only 0.1±0.5ha of farm land (CACC, 2002). These producers indicated that land size is among the main constraints for expanding their dairy farming.

Table 2: Reported land holding in the study area (Mean ±SE)

Agro-ecology				
	Highland (N=88)	Midland (N=74)	Overall (N= 162	
Land use type	Mean±SE	Mean±SE	Mean±SE	P-value
Crop land own (ha±SE)	1.01±0.62	1.34±0.64	1.16±0.65	0.001
Crop land rented (ha±SE)	0.24±0.42	0.28±0.54	0.26±0.49	0.675
Grazing land own (ha±SE)	0.35±0.22	0.30±0.22	0.32±0.22	0.199
Tree forest land (ha±SE)	0.00±0.00	0.00±0.00	0.0±0.0	0.393
Other land rented (ha±SE)	0.11±0.52	0.00±0.00	0.04±0.32	0.113

Ha=hectare, SE= standard error, HHs=households, a, b means with different superscripts for the same variable across the same row are significantly different (P<0.05).

The mean value of total local cattle breeds holding in study area was 2.4±0.3TLUand 2.6±0.3TLU per household in highland and midland respectively with the overall mean value of 2.3±0.3 TLU. Among cattle families, the largest compositions were lactating cows (30%) and followed by heifers (14.2%), bulls and male calves less than one year (10.4%), pregnant cows (7%), oxen and female calves less than one year (32%) and (4%),dry cows (2.1%) respectively per household. The cattle population in study area was significantly larger in highland agro-ecology (P<0.05) than the midland. This might be due to the agro-ecological and seasonal variation on feed availability, water, disease, land size and other factors that influence livestock keeping. From this finding it could be conclude that cattle are more dominant in highland than midland.

3.3. Milk production system in study area

In study area the dominant milk production system is mixed-crop livestock milk production system practiced by all of the rural households is (100%) in both highland and midland agro-ecologies respectively. The Pearson chi-square test indicated that there was no significant agro-ecological variation (P>0.05) about milk production system in the study area.

3.4. Major farming systems in the study district

In the study district, two major farming systems exist; crop-livestock mixed farming system and livestock production alone. In similar way, crop-livestock and livestock production systems are common farming systems in Burjidistrict of Segen Zuria Zone (Seid and Berhan, 2014).

Crop-livestock mixed farming system is a dominant livestock farming system observed in the current study area. All farmers in the study area in both agro-ecologies responded that they produce both crop and livestock simultaneously. In general, 78.9% of respondents practiced mixed production system where as 16.1% practiced only livestock production and 11.2% practiced crop production system. Similarly, 91.1% of respondent households in central Zone of Tigray, northern Ethiopia practiced mixed type of production, 6.7% livestock production and 2.2% crop production (Gebremichael *et al.*, 2015).

3.5. Reproductive performances of local and crossbreed dairy cow in the study area

3.5.1. Age at first calving

The overall estimated mean age at first calving observed in the present study for local and HF/Jersey cross breeds was 4.15 years (range 2.00 to 5.00 years) and 3.12 years (range 3 to 5 years) respectively. The result is agreed with the study conducted in the rural community of Wolaita Zone for local cows has shown that agro-ecology had no significant effect on age at first calving (Lijalem and Zereu, 2016).

3.5.2. Days open (DO)

The overall means±SE of DO reported in the present study for local and crossbred cows were 221.75±56.7 days (range 110 to 330days) and 181.5±48.90 days (120 to 340) respectively. The current value for local cows is similar with 222 days reported by Mebrahtom and Hailemichael (2016) in Endamehoni district of Tigray region.

3.5.3. Calving interval

Calving interval is the time gap between two consecutive calving. The current study (Table3) revealed that the overall mean±SE calving interval (CI) was16.15±5.11 (range 11 to 36months) and 20.01±5.53years (range 14 to 38months) for local and crossbred cows respectively. This result is similar with the report of Beriso *et al.* (2015) who reported that mean±SE of calving interval of local and cross breed cow was 19.93±0.2 and 21.7±0.3 months, respectively.

Table 3: Mean and standard error of reproduction traits in study area

| | Reproductive traits | | | | |
| Variable | Highland | Midland | Total | P-Value | |
	Mean±SE	Mean±SE	Mean±SE	Agro	Breed
AFC: Lbr	4.2±0.05	4.1±0.05	4.2±0.04	0.209	0.000
AFC: Cbr	3.0±0.06	3.1±0.05	3.1±0.04		
DO: Lbr	202.13±5.8	239.6±5.9	219.9±4.45	0.915	0.000
DO: Cbr	185±14.4	144.7±21.3	172.±17.9		
CI: Lbr	14.4±0.54	18.8±1.29	16.15±0.42	0.036	0.012
CI: Cbr	18.2±0.57	20±1.89	19.2±1.23		

SE=standard error, Agro=Agroecology, Lbr=Local breed, Cbr=Cross breed

3.6. Productive performances

3.6.1. Daily milk yield

The average milk off-take of local cows was about 1.24±0.02 litre/head/day (Table 4). As indicated in Table 4, the overall mean of local and crossbreed cow milk yield/head/day at first, second, and third lactation stages was 2.23±2.85 litre, 1.52±2.28 litre and 0.72±1.23litre, and 4.53±0.621, 7.06±1.72 and 4.43±0.85 respectively.

3.6.2. Lactation length

The overall average means lactation length of local and crossbreed cows observed in the current study were 8±0.19 months and 9±0.57 months respectively.

The current study result is in lined with the reports of Beyissa *et al.* (2017), Mebrahtom and Hailemichael (2016); Melku *et al.* (2017), and Tadesse and Tadelle (2003), indicated the lactation length for local cows of 7.58, 8.93months, 8.68 months and 9.34months respectively.

3.6.3. Dry-off period

The overall average means dry period length of both local and crossbreed cows observed in the current study were 66.9±17.77 days. This result is in lined with the report of Atashi*et al.* (2013) who reported that mean dry period length was 67.3 (23.7%) in Iran. The result of current study dry period length was lower than the report of Sandra *et al.* (2014) only one farm (1.1%) had a regular dry period of more than 70 days dry period length in northern Germany.

Table 4: Mean and standard error of production traits in the study area

Variable	Production traits				
	Highland	**Midland**	**Total**	**P-Value**	
	Mean±SE	Mean±SE	Mean±SE	Agro	Breed
DmyEL:Lbr	1.6±0.1	1.8±0.11	1.7±0.11	0.015	0.000
DmyML:Lbr	2.0±0.3	2.1±0.2	2.1±0.3		
DmyLL:Lbr	0.5±0.1	0.6±0.1	0.6±0.1		
DmyEL:Cbr	7.4±0.3	6±0.4	6.7±0.35		
DmyML:br	6.8±0.4	10.9±0.6	8.9±0.5		
DmyLaL:Cbr	2.9±0.1	3.8±0.2	3.4±0.2		
LL:Lbr	7.8±0.2				
LL:Cbr	8.3±0.5	8.3±0.2	8±0.2	0.02	0.02
DPL:L/Crbr	67±9.6				
		9.2±0.7	9±0.6		
		67±23	67±16.3	0.883	

SE=standard error, Agro=Agroecology, Lbr=Local breed, Cbr=Cross breed, Dmy=daily milk yield, EL=early lactation, ML=mid lactation, LaL=late lactation, LL=lactation length

3.7. Milk and Product Marketing in the Study Area

3.7.1 Involvement of producers in dairy marketing

In the study district (Table 5) , the majority (57.9%) of dairy farmers produced butter as the predominant dairy product for sale while 38.0% of households produced sour butter milk for sale and 4.1% of households sold whole milk and the rest sold cottage cheese and ergo. In this district, the amount of income from the sale of whole milk was low. The major dairy products used for income generation in this area were only butter and sour buttermilk and cheese. Out of the total sour buttermilk produced after churning, a higher proportion was used for household consumption while the rest was sold.

Similarly, Nigussie (2006) reported the absence of formal marketing system in Mekelle urban dairy system. In contrast, because of the presence of milk processing plants in Addis Ababa there are emerging formal marketing systems in the Addis Ababa milkshed (Sintayehu *et al*, 2003).

The majority of the women in the study district sold their dairy products (butter and cheese) once a week (during large market day) and sold buttermilk at the time of processed day. The majority of the farmers are living at about one hour walking distance from the local market place and there is informal dairy marketing system in place. The small quantity of butter produced per day may not be accumulated to the desired volume to be marketed. On average they travel about 5 km (ranging from 1 to 10 km) to market places.

About 75.2% of the respondents travel more than 10 km in a single trip on foot and few (24.8%) respondents travel by cart to take their products to market. Milk products were marketed on the basis of volume rather than weight. Butter was commonly measured in clay cups, which have different sizes (small, medium and large).

Two medium clay cups were equivalent to one kg of butter and sold at the rate of 173.40 Birr/kg of fresh butter. Butter was also sold by putting on enset (false banana) leaf of different sizes that is measured by fingers. On the average five fingers` butter were equivalent to one kilogram of butter. These estimations were based on experience and mostly decided by butter traders and producers. The middlemen collect most of the benefits by buying from the rural producers on the basis of rough estimates of volume by balancing his/her hands up and down throwing of butter and selling on the basis of weight to consumers in urban areas. In the study areas, there was no report of marketing sour milk, whey and nitirkibe (ghee) because they prepare ghee for their consumption.

In the study area, the quality parameters of milk and milk products are cleanness (32.3%), colour (3.1%), freshness (8.1%), and flavour and aroma (36%). As interviewed producers told as, the buyers look the moisture content of butter and cheese, and if there were high moisturity, they purchase in low price.

Table 5: Preconditions for dairy products marketing in study area

Parameters	Agro-ecology		
	Highland	Midland	P-value
Collecting butter & cheese for sale	%	%	
Yes	95.4	93.2	0.552
No	4.6	6.8	
Means of transportation			
On foot	83.9	64.9	0.005
By cart	16.1	35.1	
Quali. Factors for sale &consum.			
Cleanness	41.4	21.6	
Colour	0	6.8	
Freshness	11.5	4.1	0.002
Flavour and aroma	28.7	44.6	
All	18.4	23.0	

N=Number, Quali=quality,consum= consumption

3.7.2 Marketable dairy derivatives and prices

The average daily milk off-take of local cows was about 1.0 litre/head/day. On average about 20.96±32.13 litres of milk was produced/household/week out of which about 12.53 litres was accumulated for further processing and the remaining 8.43 litres consumed on weekly basis. On average about 1.12 kg of butter was produced per household per week in both the two agro-ecologies. Out of this total production about 1.12 kg per week (85%) was marketed and the remaining for home consumption. Majority of women interviewed in study district made cottage cheese. Majority of respondents who were produce cottage cheese use both for home consumption and sale. On average about 5.59 kg of cheese was produced per household per week in both the two agro-ecologies. Out of this total production about 1.12 kg per week (85%) was marketed and the remaining for home consumption.

Majority of the respondents in highland and midland did not sell fresh milk due to insufficient production and cultural taboo but few respondents sell fresh milk that have cross breed cow only. There is a need to strengthen extension activities to increase milk production in the area and to change the attitude of farmers toward fresh milk sale. The establishment of organized milk collection and marketing infrastructures would encourage them to change these trends.

3.7.3 Determinants of price, demand and supply of milk products

During the survey period (January), the average price of buttermilk (``arera``) of local cow and fresh milk of crossbreed cow in the study area was 10.00 ETB per litre and 25 ETB respectively. The average price of butter was 190.5 ETB per kg with a minimum and maximum price of 173 ETB and 208 ETB per kg, respectively. The average price of cheese was 43.5 ETB per kg with a minimum and maximum price of 36 ETB and 51 ETB per three up to five days` collection, respectively.

From this study it was noted that various factors affect the price, demand and supply of milk products in the study area. These included season (dry versus wet), distance to market points, fasting periods, festival and holidays. The results of this study are similar to the findings of (Sintayehu *et al.*, 2008). During the wet season due to better availability of feeds there is an increase in milk yield and in turn other milk products per household and per animal compared to the dry season, hence, the better supply to the destination market.

3.7.4 Butter trade routes and trading activities

In the study area, inflow of butter into the studied areas from Kucha, Dawuro, KindoDidaye, Ofa and Arba-minch areas was transported long distances of over 170km up to Hawassa and Shashemene. Although butter from this area, particularly from Soddozuria district kebeles was carried long distances. Therefore, butter is the most marketable dairy product having more market channels over long distances.

3.8. Milk production constraints in the study district

The four major constraints that were ranked from first up to fourth by the respondents during individual interview. These were low amount of fodder or roughage, low milk yield of local cows, low quality of fodder or roughage and high price of cattle feed (concentrate) but some others were shortage of feed, shortage of land, disease outbreak, shortage of labour, and absence of market from other areas for products. The respondents were asked to rank these problems and lack of fodder or roughage has got the highest rank in both agro-ecologies. Next to lack of fodder or roughage, most study households (64.9%) ranked low milk yield as another milk production constraint in highland and low quality of fodder or roughage in midland, this agro-ecological difference was due to number cows per household and breed type of dairy cows. Most study households (87.2%) ranked high price of cattle feed (concentrate) the fourth milk production constraint in the study area.

3.9. Milk and milk products marketing constraints

The milk and milk products marketing system in the study area face severe constraints in both agro-ecologies was (83.9%). The major constraints for milk and milk products marketing identified by the farmers were low price of milk (77.4%), poor quality of milk/sour milk (13.7%), and lack of transportation (5.5%), no market (2.1%) and others (1.4%) such as high cost of transport, long distance to market, low products and lack of milk preservation methods for buttermilk and cottage cheese. Cultural taboo is indicated as a limiting factor of males for milk and milk products marketing that done by females only. This result is contrary to the report of Tegegne *et al.* (2013) in east Shewa Zone of Oromia, that among the many reasons reported by farmers, insufficient amount of milk production and cultural restriction were the most common hindering factors.

Majority (88.8%) of respondents reported that there was no delay in getting paid for milk sold, due to low milk yield and few (11.2%) respondents had a delay in getting paid for milk sold. In the study district, majority (93.2%) of farmer responded that there is no spoilage of milk due to lack of market and few (6.8%) producers responded that there is spoilage of milk due to lack of market.

As indicated by producer respondents, from the total respondents, days of butter collected /household/week were 4-5 days and 3-4 days (cheese) of the respondents respectively and frequency of selling milk/month were 16 times that is not stored due to rapidly fermentation and spoilage and to buy daily needs of salt, coffee-beans, breakfast foods (maize, pea, bean). In generally, the dairy producers reported that they store especially processed products (butter and ghee) from 2 weeks-4 years before selling.

3.10. Opportunities for milk and milk products developments

Although many problems and constraints that may hamper the development of the dairy sector were identified in the study district, the majority of dairy producers of both agro-ecologies the highland (94.3%) and midland (85.1%) production systems were willing to continue, expand and/or involve in dairying in the future. The rest of the producers were not willing to expand dairying in the future for various reasons. About 5.7% and 14.9% of the respondents in the highland and midland, respectively, indicated that they will stop dairying, respectively.

Generally the highland producers were more willing to continue and expand dairying due to nearest to Soddo town and market opportunities in areas. Because of the rapid urbanization of

Soddo town, substantial population growth and change in the living standard by highland societies in the area, the demand for good quality and quantity of dairy products are increasing.

Dairying provides the opportunity for smallholder farmers to use land, labor and feed resources and generate regular income. Although market opportunity and linkage are key issues for smallholder dairy development, support services in terms of accessing adequate land, organizing input supplies (improved genetic material, feeds, AI, drugs), provision of credit, extension and training services, production and entrepreneurial skills developmentare key elements for success (Sintayehu *et al.*, 2008).

4. CONCLUSION AND RECOMMENDATION

The major milk production system practiced in the study area was crop livestock mixcd production system identified and characterized. Most of the foundation stocks of both the highland and midland producers were purchased from open markets, which revealed that farmers were not curious and/or did not have access to the selection of dairy cattle. Reproductive traits of AFC and day's open of local cows was longer as compared with crossbreed cows, however calving interval of crossbreed cows more than local cows.

Marketable milk and milk products in the study areas were butter, buttermilk, cheese and fresh milk rarely. The dairy marketing system identified in the study area was entirely informal marketing system, in which the farmers sell milk and milk products directly to consumers and traders. The most common butter marketing channels were from the producers to consumers, producers to traders and from traders to consumers.

In general, milk and milk products marketing system were constrained by breed of cattle, lack of milk processing services and equipment, poor market information on the price and supply condition, lack of services (extension, inputs, and veterinary) and lack of feed type, feed processing and utilization management. Therefore, to improve the situation, use of better feed conservation and utilization techniques, use of improved feeding system and improved animal health services are believed to solve these problems. In order to achieve these, provision of training to the farming communities is imperative so as to improve their knowledge and skills on the management of dairy and beef animals. They also should be reinforced through increasing dairy market out lets by forming market oriented dairy producer and marketing cooperatives and improving infrastructure facilities in order to reduce transaction cost related with distance from milk market out lets and market evidence.

5. REFERENCES

Abraha Negash. 2018. Ethiopian Meat and Dairy Industry Development Institute, Ethiopia.

Addis Getu, Tadese Guadu, Shewangizaw Addisu, Asechalew Asefa, Maleda Birhan, Nibrete Mogese, Mersha Chanie, Basazenew Bogale, Atnaf Alebie, Atsede woyne Feresebhate, Tegegn Fantahun and Tadegegne Mitiku. 2016. Cross breeding challenges and its effect on dairy cattle performances in Amhara region, Ethiopia. Online J. Anim. Feed Res., 6(5): 96-102.

Agajie T, Chilto Y, Mengistu A, Elias Z and Aster Y. (2002), Smallholder livestock production systems and constraints in the highlands of North and West Shewa Zone. Ethiopian society of animal production (ESAP) proceeding of the 9th annual conference held in Addis Ababa, Ethiopia, August 3031, 2001. PP: 49-71.

Ahmed M., Ehui S. and YemesrachAssefa. 2003. Milk development in Ethiopia. Socio economics and policy research.Working paper 58. ILRI (International Livestock Research Institute), Nairobi, Kenya. 47p.

Ahmed M., Ehui S. and Yemesrach Assefa. 2004. Dairy development in Ethiopia. EPTD Discussion Paper No. 123. Washington DC: International Food Policy Research Institute.

Asrat Ayza, ZelalemYilma and Ajebu Nurfeta. 2013. Characterization of milk production systems in and around Boditti, South Ethiopia. Livestock Research for Rural Development. 25(10).

AsratAyza, Zelalem Yilma and Ajebu Nurfeta. 2014. Production, utilization and marketing of milk and milk products: Quality of fresh whole milk produced in and around Boditti, Wolaita, and South Ethiopia. LAP, LAMBERT Academic publishing, Deutschland, Germany.

Atashi H., Zamiri MJ., Dadpasand M. 2013. Association between dry period length and lactation performance, lactation curve, calf birth weight, and dystocia in Holstein dairy cows in Iran. J Dairy Sci. 2013 Jun;96 (6):3632-8.

Bayisa Amenu, Ulfina Galmessa, Lemma Fita and Belay Regasa. 2017. Assessment of Productive and Reproductive Performance of Dairy Cows in Gindeberet and Abuna Gindeberet Districts of West Shoa Zone, Oromia Regional State, Ethiopia. Journal of Biology, Agriculture and Healthcare. 7(10): 60–65.

CACC (Central Agricultural Census Commission). 2002. Ethiopian agricultural sample enumeration report, held 2001/02 (1994 EC).1990. ILCA (International Livestock Centre for Africa), Addis Ababa, Ethiopia.

CSA (Central Statistics Authority). 20017/18. Agricultural sample survey 2017/2018. Vol. II. Report on livestock and livestock characteristics. Statistical Bulletin 578. Addis Ababa, Ethiopia: CSA.

FAO. 2004. Livestock Sector Brief in Ethiopia, Food and Agriculture Organization of the United Nations. Livestock Information, Sector analysis and Policy Branch.

Gebreegziabher Zereu and TsegayLijalem. 2016. Production and Reproduction Performances of Local Dairy Cattle: In the Case of Rural Community of Wolaita Zone, Southern Ethiopia. Journal Fisheries Livestock Production. 4:176.

Gebremichael. D, Belay.B, Tegegne.A (2015). Assessment of Breeding Practice of Dairy Cattle in Central Zone of Tigray, Northern Ethiopia',Jornal of biology,agriculture and healthcare vol 5(23), pp. 96–105.

Getachew, M. and Tadele, Y. 2015. Constraints and opportunities of dairy cattle production in Chencha and Kucha districts, Southern Ethiopia. Journal of Biology, Agriculture and Healthcare, 5(15), pp. 38 43.

Getu A, Guadu T, Addisu Sh, Asefa A, Birhan M, Mogese N, Chanie M, Bogale B, Alebie A, Feresebhate A, Fantahun T, Mitiku T. 2016. Crossbreeding challenges and its effect on dairy cattle performances in Amhara region, Ethiopia. Online J. Anim. Feed Res., 6(5): 96-102.

Haese M D, Francesconi G N and Ruben R. 2007. Network management for dairy productivity and quality in Ethiopia Development Economics Group, Wageningen University, Hollandseweg1, 6706KN, Wageningen, the Netherlands. Holdings).Volume V. Addis Ababa, Ethiopia.

Ketema, H and Redda T. (2004), Dairy production system in Ethiopia. Ministry of Agriculture Ethiopia. Pp1.

Ketema S. 2008. Characterization of market oriented smallholder dairying and performance evaluation of dairy cooperatives in TiyoWoreda, Arsi Zone of Oromia Regional State. Msc Thesis. Hawassa University, Hawassa, Ethiopia.

Kumsa T. (2002), Smallholder dairy in Ethiopia. Bako Agricultural Research Centre P.O. Box 3, Bako, Ethiopia.

Mebrahtom Bisrat and Hailemichael Nigussie. 2016. Comparative Evaluation on Productive and Reproductive Performance of Indigenous and Crossbred Dairy Cow Managed under Smallholder Farmers in Endamehoni District, Tigray, Ethiopia. Journal of Biology, Agriculture and Healthcare. 6(17).

Melku Muluye, KefyalewAlemayehu and Solomon Gizaw. 2017. Milk Production Performances of Local and Crossbred Dairy Cows in West Gojam Zone, Amhara Region, Ethiopia. Journal of Applied Animal Science. 10(1): 35-46.

Nigussie Gebreselassie. 2006. Characterization and evaluation of urban dairy production system in Mekelle city, Tigray region, Ethiopia. MSc thesis, Hawassa University, Awassa, Ethiopia. 54 pp.

O'Connor C.B. 1990. Rural smallholder milk production and utilization and the future for dairy development in Ethiopia. Dairy marketing in sub-Saharan Africa Proceedings of a symposium held at ILCA, Addis Ababa Ethiopia, and 26-30 November 1990. International Livestock Centre for Africa P.O. Box 5689, Addis Ababa, Ethiopia. Pp126-133.

Prain G, Karasna N and Smith D. 2010. African union harvest agriculture in the cities of Cameron, Kenya and Uganda. International development research centre PP335.

Redda T. (2001), Small scale milk marketing and processing in Ethiopia. In proceedings of south work shop on small holder dairy production and marketing, constraints and opportunities. March 12th-16th. Anand, India.

RLDC 2009. Rural livelihood Development Company. Dairy sub sector in sub Saharan Africa report of work shop held on 3-5 march 2003 in Nairobi Kenya. Natural resource Ltd, aytesfordkemt, uk, pp118.

Sandra Bertulat, Carola Fischer-Tenhagen, Wolfgang Heuwieser. 2014. A survey of drying-off practices on commercial dairy farms in northern Germany and a comparison to science based recommendations.

Seid Guyo Guje and Berhan Tamir. 2014. Assessment of cattle husbandry practices in Burji Woreda, Segen zuria of SNNPRS, Ethiopia. International journal of technology enhancements and emerging engineering research. 2(4).

Sintayehu Gebre Mariam. 2003. Historical development of systematic marketing of livestock and livestock products in Ethiopia. In: Jobre Y and Gebru G (eds), Challenges and opportunities of livestock marketing in Ethiopia. Proceedings of the 10th annual conference of the Ethiopian Society of Animal Production (ESAP) held in Addis Ababa, Ethiopia, 22–24 August 2002. ESAP, Addis Ababa, Ethiopia. pp. 15–21.

Sintayehu Yigrem, Fekadu Beyene, Azage Tegegne, and Berhanu Gebremedhin. 2008. Dairy production, processing and marketing systems of Shashemene–Dilla area, South Ethiopia pg-32. IPMS (Improving Productivity and Market Success) of Ethiopian

Farmers Project Working Paper 9.International Livestock Research Institute, Nairobi, Kenya. 62 pp.

SZWAO (Soddo Zuria Woreda Agricultural Office). 2009. Rural and Agriculture Department of Soddo Zuria Woreda, number of crop plants annual reports.

SZWLF (Soddo Zuria district Livestock and Fishery Office). 2010. Rural and Agriculture Department of Soddo Zuria Woreda, number of livestock annual reports.

SZWLFO (Soddo Zuria Woreda Agricultural office). 2011. Rural and Agriculture Department of SoddoZuriaWoreda, annual reports.

Tadesse Million and Tadelle Dessie. 2003. Milk production performance of Zebu, Holstein Friesian and their crosses in Ethiopia. Livestock Research for Rural Development. 3 (15).

Tegegne, A., Gebremedhin, B., Hoekstra, D., Belay, B. and Mekasha, Y. 2013. Smallholder dairy production and marketing systems in Ethiopia: IPMS experiences and opportunities for market-oriented development. IPMS Working Paper 31. Nairobi, Kenya: ILRI (International Livestock Research Institute).

Tollossa Worku, Edessa Negera Gobena, Ajebu Nurfeta, and Haile Welearegay. 2014. Milk handling practices and its challenges in Borana Pastoral Community, Ethiopia. African Journal of Agricultural Research.9(15):1192-1199.

Weldegebriel, D. G. 2015.Assessment of production and reproductive performances of cattle and husbandry practices in Bench-Maji zone, Southwest Ethiopia, Global Journal of Animal Scientific Research. 3(2): 441-452.

Disclosure: None of the authors have any conflict of interest.

Acknowledgement

The authors greatly thank the smallholder farmers of Sodo Zuria districts, traders and others stockholders for their cooperation in providing necessary information. The financial support of Wolaita Sodo ATVET College is gratefully acknowledged.